THE AIR SHOW

Also by Peter Scupham:

Prehistories (1975)
The Hinterland (1977)
Summer Palaces (1980)
Winter Quarters (1983)
Out Late (1986)

THE AIR SHOW

Peter Scupham

Oxford New York

OXFORD UNIVERSITY PRESS

1988

Oxford University Press, Walton Street, Oxford OX2 6DP

Oxford New York Toronto
Delhi Bombay Calcutta Madras Karachi
Petaling Jaya Singapore Hong Kong Tokyo
Nairobi Dar es Salaam Cape Town
Melbourne Auckland

and associated companies in
Berlin Ibadan

Oxford is a trade mark of Oxford University Press

© Peter Scupham 1988

First published 1988

British Library Cataloguing in Publication Data
Scupham, Peter, 1933–
The air show.—(Oxford poets).
I. Title
821'.914
ISBN 0–19–282206–3

Library of Congress Cataloging in Publication Data
Scupham, Peter, 1933–
The air show/Peter Scupham.
p. cm.—(Oxford poets)
I. Title. II. Series.
PR6069.C9A77 1988
821'.914—dc19 88–1514
ISBN 0–19–282206–3

Set by Wyvern Typesetting Ltd.
Printed in Great Britain by
J. W. Arrowsmith Ltd., Bristol

For Ann, who was there too

Acknowledgements

Some of these poems first appeared in *Christmas Visits*, (Mandeville), *Encounter, The Literary Review, Outposts, PN Review, The Poetry Review, Strawberry Fare, The Woodbury Review*, and *Words International*, or were read on *Poetry Now* (BBC 3).

'Neighbour', 'The Spanish Train' and 'Pathfinder' were published in *Summer Palaces* and *Winter Quarters* (OUP). A selection of these poems has been published in a limited edition titled *Under the Barrage* by Hans van Eijk, In de Bonnefant, Holland.

Contents

Toy Chest

With a slow, makeshift groan
   The toy chest swings its lid,
Nodding dull rafters
   Over stuff hid.

I trace the knotted wood,
   The paint's untroubled skin,
Toy chest; blanket box—
   Their rough squat settles in

Folding like summer nights
   Over dead games,
The struck and puzzling hours,
   The bright, unwanted names.

There, jammed in the cracks,
   A smell of other rooms:
Young, sharp voices,
   Quick tear-storms.

Those years that pulled me by,
   Caught in their apron-strings,
Crouch in odd corners
   With hugger-mugger things

Black, lumpy as sacks
   Worked-up by Burke and Hare.
Blindfold, I guess the weight
   Of all that's gathered there

Refusing to come out
   In this pale April light:
Tongueless, hot, obdurate,
   Packed in too tight.

BACK

Back

I drew the car up in the land of was;
It did not seem to mind the different air,
But shone, its dials playing fast and loose,
The mileage clocking up another year.

I recognized my voice. The gates had gone,
And someone odd was snicking banks of green,
Softening the path down which my footsteps ran
With softer shadows. I was back again.

The house was smaller, shrunken on the bone;
He turned and walked through what I could not find.
I was his future for an hour, he mine;
We watched each other printing out the wind

Which slipped across to nowhere as we smiled,
Huge cloudy summers piling up behind.
I praised the shipwrecked garden he had filled
With stupid, alien stuff, and stood my ground

Where shining levels, ordered beds, a sun
Immovable and steady in its fire
Let all his pleasant gossip trundle on
Past savaged trees, the dense and loaded air.

'But tell me, was it haunted? There's a ghost . . .'
I laughed: 'Unless it's me . . .' and stared past him
To where some prying fiction of my past
Inked out the shadows in that upstairs room.

'Come in, and have some tea.' I would not go:
A summer child, out in the garden late,
Who knew that in means talk, not much to do,
Just sitting, watching grey things draw to night.

The Air Show

I think I can see myself—a small animal
Doing a hop-dance over a patch in my head
Which is called summer and grass; where light is cut
Into bits and ribbons. I name this space The Air Show,
And most of it now is perfectly dismantled:
The whirring, the whining, the scratches laid on the air
Spin from a gramophone's headpiece rocking its needle
Over and over into those neat ellipses.
The song has run itself out at the end of the line
Into the blackout, past the unreadable label,
That frog-voice gone which croaked so briskly
'In a Quaint Old Normandy Town', or groaned in vain
For 'Amy, Wonderful Amy'. The summer is thick,
Here, in Grandad's Office. I dowse the flies with Flit,
Gunning them down upon the dazzling window,
Watching them buzz their clockwork, kick up their legs,
Stiffen into burnt raisins, crumbs, dustiness,
As their stuck-up friends drown in yellow flypaper.

At The Air Show, beyond the bobbing heads,
The buzz of flies and voices, bits and ribbons,
A tri-plane bellies its fabric into the sunshine,
Wing shelved on wing, engine shelved on engine:
Stuck monstrously on the grass. High overhead,
The little silver gnats tow rolling banners
Lettered with huge but cryptic messages.
The sky is very open, very endless,
And lying on my back with my head cupped
I feel the world go twirling softly round
Like an old waltz: 'Destiny', 'Romola',
While my mother in white dances into the 'thirties.
These deeps are blue, the stuck specks in my eyes
Are looping loops in fields of vertigo,
And that prodigious, glittering machine
Rests delicately on a sheet of emerald,
Being itself, The Air Show, gathering light
As the years grow smaller, tucking themselves away
Under the steepling ledges of its wings.

The Path into Avernus

The Christmas Grotto lives inside the tent
Of darkness childhood pitched inside my head,
Where someone else, whose hair is neatly parted,
Is walking hand-in-hand, and with the dead.

The airs are green, the path into Avernus
Wavers through light which bulges into rock,
Against the heigh-ho's and the cardboard larches
The seven dwarfs are working round the clock,

Swinging their axes into wooden billets,
And goggling back with huge aquarium eyes
At Innocence, her white dress falling, falling
Softly as snow into her own surprise.

Along the golden mile, illuminations:
The golden road, and far-off Samarkand.
Under preserving ice in Xanadu
The child, the dead are walking hand-in-hand.

The Spanish Train

The little Spanish train curls in the hand,
Its coat of many colours; the June garden
Will blow to seed, find new snows and sierras.
Somewhere, beyond the phlox, the cherry wall,
Goya etches: *tristes presentimientos.*
And though his rooms are hung with all misfortune
The train draws down a truthful patch of sunlight,
A radiance not yet underpinned by shade
Or lost in the earth-closets of the garden:
Neither the last mile to Huesca taken,
Nor the fixed siren set upon the Stuka.

Now, on a sofa, the child holds a word:
Spain, where the rain goes, and a wooden train
Quite serious in its unclouded paintwork,
Its yellow bright as any Star of David.
While the small fingers look inside a carriage
Or hook and eye the polished dolly-waggons
The flowers prepare their faces for the night.
Shutters and bolts are drawn. There are long journeys
Which must be made. *No saben el camino.*
There is no remedy. There is no time.
The little Spanish train curls in the hand.

1 Crich Circle, Littleover, Derby

I'd like to pull you back into the sun
That beats the boundaries of your patch of lawn,
Your swags of elder, poplars, privet-hedge,
But you refuse: unable to dislodge
The load of night which presses on your roof,
That crooked, dismal, half-enchanted life
You share with an abandoned garden shelter,
Old corrugated iron and black water,
The colonizing rats who scamper still
Up the forsythia pinned back to the wall,
The tamped and piggy sandbags swollen thick
Against french windows always on the sneck.
You like to keep your smoky head well down,
Bunkered against your neighbours, but alone,
As if night had to fall, and fall again
While you, the trap, coax all the darkness in,
The Deco flowers blazoned on your glass
No prophylactic against dream and loss.
You are the ache, the hug and cloudiness
Of ghost-life pressed and whispering between
The day's undressing and a greyish dawn,
Cold knobs and corners stubbed against my head:
A labyrinth, and now my only guide
This pocket torch whose battery runs down
And dwindles to a fleck of yellow shine,
A tiny frizz of wire which focuses
On a child's book where all the small print dances
Out of my reach. You drift into the moon:
A twist of shrapnel, silver at the skin,
Coldish and rough and crumbly further in.
A humming-top, you spin the bombers on
To one deep-throated and unsteady drone.
Between the first Alerts, the last All-Clears,
I cannot place you: one of ours, or theirs?

Diet

They gave me bread and brains and buttered porridge,
A dusty glass of water, stale and sweet,
To take away the sickness of the roses
That climbed the bedroom walls and could repeat
Nothing but roses, roses, till their nonsense
Crept sweating round me in a winding sheet.

They pushed their gravy trains into my tunnel
And stuffed me up with garden slugs and snails.
My salad days were crisp and limp as lettuce,
Then all the clockwork toppled off the rails,
The little foxes jumped from all the foxholes
And ran about the world with blazing tails.

They rubbed their hands in sorrow and in anger,
Then put the crumbs out on the window-ledge:
The crows preferred to gobble up the eyeballs;
And all the children's teeth were set on edge.
The devil spat on all the plumpest brambles
And sharpened up his nails on every hedge.

They laid the tables with their best behaviour,
And stuffed my manners in a napkin-ring.
I told their boasting turkey it was Christmas
And sang that I was happy as a king.
Then when it lay in bits of skin and rubble,
I cracked its wishbone, wished like anything.

They never thought, when wishes turned to horses,
And all the beggars rode away, a set
Of crows and foxes, eyes as big as saucers,
Would fight for any sweetmeats they could get.
I egg them on by eating to remember;
I buy them off by drinking to forget.

Double

She says he came back that night. She lay awake,
The usual skeletons propped up in their cupboards
With best-china faces. The guns began to bark;
The shrapnel pattered. When more stones fall than usual,
Consult the Sibylline Books. I do consult them,
But the pale snapshots in their padded albums
Are ancient lights, ambiguous oracles.
We know, though, he was out, and at his post:
One of those Augurs in their stencilled helmets
Reading the scrawled sky, the city's entrails.
Stoics and bombs whistle as lightning strikes;
The skeletons crooned, bouncing up at keyholes.
She, hoping, praying that we would make old bones,
And, as she swears, with all her wits about her,
Turned to find him there in the smallest hours,
Moving about the room he had shored up
With baulks of timber. Under those auspices
They chatted together for a little while,
Her fear draining away, the heartbeats gentle,
Before he closed the door upon himself,
Blacked out his shadow on that darker shadow
So tightly they grew indivisible.

Later, much later, she got out of bed
And laid a breakfast of crossed purposes,
While he, rubbing the fires from his eyes,
Wrote out a testament, alas, gone missing,
To clear himself of presence without leave
That night when Love and Duty had his name
So clearly down upon their distant rosters.
Martial says 'Dream of yourself, or stay awake.'
She says that he came back; he still denies it.
I was asleep. What I saw I will not tell you.

A House of Dolls

Someone who spilt salt, walked under ladders,
Has wished bad luck on Mr and Mrs Brown.
Who cracked the mirror? Who sent a letter
And stammered the King's head on upside down?

Their smudgy faces have been cried upon—
We will not say by whom—and though they tell
Make-believes about cats in the lonely evenings,
The mainspring is broken: nothing is, goes well.

Someone took pennies from a someone's handbag
And was shut in a high room till the devil went;
The car lay on its back with its wheels kicking,
Chocolate money grew scarce, and was soon spent.

It was scrimp and save from now on for the duration,
The breakfast scraps tacked to the dinner-plates,
The children wriggly and skinny as young rats,
The calendar a migraine of blind dates.

Whose tongue and lips on the battery's terminals
Felt the juice tingle as it ebbed away,
Pulling the light out of the clickety windows?
God found somewhere else where he wanted to play,

Bored with banging life shut or swinging it open,
The doll's house bared like a split face
On sobs and dust and bits and wet corners,
Something missing, and nothing in its place

But the Brown family, mended and making-do,
Propping their pipe-cleaner headaches against the wall,
All the invisible stuffing knocked clean out of them,
And nowhere to go to get away from it all.

The Old Frighteners

When the impossible grottoes hunch their backs
Under the crinkled stars, we dream to find
The sirens blowing down the bedroom chimney,
Far, far away, and paler than the wind

On darkness hanging from the window-panes.
A pail of water and a pail of sand
Will keep the devil in, and devils out.
What is it that goes round the house, and round,

Dragging its tarnished ribbon off the spool,
Creasing its wicked face upon the moon?
Unsteady things call to unsteady things;
All the slack fur is rubbed against the grain.

Whoo-hoo, whoo-hoo go the old frighteners,
Their clammy sheets twisting between our toes,
Banging the doors about in the grey ghost-house,
Pushing their long dull fingers into eyes

Sticky with sleep, half-opening to find
The sad lights fluttering their yellow-brown
Off the soft edges of the bannisters.
We float downstairs upon the ache and drain,

The unison, the dying aah of things
Traipsing their nonsense back into the grave.
Then the bristle of night standing on end:
The guns chuckling out in the witch's grove.

Children of Odin

'Once there was another sun and another moon.'
They shine through the pages of a battered book
Where I and Heimdall keep the Rainbow Bridge,
Hearing the grass grow, watching a hundred miles
Over these branching streets, dark snow, lit trees
And stars silverily adrift over the houses.
In Christmas basements hairy Brock and Sindri
Forge me Skidbladnir, my pocket-battleship,
Scuttling itself about in soapy waters.
Like them, I know the underside of things,
The brown roof of a table, drapes, rock walls
Which echo to Thor's hammer, huge Miölnir
Storming its way down gloomy roads to nowhere.
I must be Ratatösk, the Squirrel of Mischief,
Slipped between the worlds of the dead and living,
Between the seconds of the front-room clock,
Between the grains of sand which pad the sandbags
Layered like drowsy pigs against the window.
I rehearse the names proper to my fable:
Hugin and Mugin, perched on Odin's shoulders,
Riding the night with Junkers, Dornier.
Hitler flings his right arm at the air
And Baldur drops, struck dead by mistletoe
Whose waxy clusters drip over the doorway:
Dull shrouds for seedy candles, oily lanterns
Lighting the forest acres and more snow.
I will lie on the Fields of the Dead, Hela my bride;
Half-corpse, half-woman—she will gather me
Out of the daylit world into the flight of things
Riding in from the East, over the North Sea.
Behind that other sun and other moon, wolves run.
Their jaws are open; they will eat the light,
Leave me for ever in the dark, the cold.

Jungle Book

Cold Lairs: the sheets are fighting like ghosts tonight
Under an eiderdown prinked out with stars,
Brilliant stars cut from a wambly night-light
Shiftlessly trembling into its pool of wax,
Sheathed in a ring of card. The Bandar-log
Swing fast from dream to dream, their jungle-stuff
Twitching the shadows out of secret corners.
I live under the white of the lost Queen's dome,
Her palaces of pleasure oiled by darkness.
At my feet, a stone water-bottle
Nudges its blunt head, ominous as Kaa's;
Under the bed the nestling cobras pry,
Swelling their black hoods, readying their fangs,
And lawless beasts are banging about outside
Where reservoirs and memories crimp the moon,
Dowsing the shaky night and all its fires,
Turning its bones to water. In the cities
There will be burning, bodies under blankets,
Wallpaper dangling like wet skin from a scald—
Here, minarets and lattice-work in stone
Under the night-sky of an eiderdown.
Heavy Baloo and lemon-eyed Bagheera,
Warm your kind fur out of the bedclothes' ruck;
It is Cold Lairs: I know the call-sign,
'We be of one blood, ye and I. . . .'

Glory

Things lie on their sides, mainsprings uncurling,
Kicking their wheels up, buzzing with awful rumours
To dandelion-clocks and stumpy daisies.
The garden spreads its nets of camouflage
For *Britain's Wonderful Fighting Forces*; here,
The life of insects and the life of grasses
Is smudged aside by our advancing armies,
The caterpillar-tracks, shed tyres, jumped-out-of skins.
The *Men of Steel* leap from their photographs
Into this humming, fluttering summer cauldron,
Commandeering chipped green dinky-wagons,
Their heads propped high on pride and broken matchsticks.
Held in reserve *Our Unchallengeable Navy*
Lurks in Davy Jones' thick-lidded toybox,
Its thin grey line patrolling those dark emperies,
And somewhere night's black exo-skeletons
Wait to pounce with their cold magic claws
On all our silver *Warriors of the Skies*
Drawn up in squadron at the lawn's far corner.
I watch my veterans with their squared-off faces
Lidded with steel like ancient Chinamen,
Moving in quick slow-motion through a smokescreen
Of high, impossible words and windy leaves.
When our campaigns dwindle and die at sunset,
The Children's Encyclopedia calls *Make and Do*,
Opening my cats' eyes to its *Book of Wonder*,
Its *Magic String*, index of exclamations.
All these words pump up a special glory
As hard, unpinchable as my bicycle tyres.
I stick a fan of paper between the spokes
And soldiers, history, I, my garden-friends,
Race like wound-up motors into our futures.

JOURNEYING BOY

Journeying Boy

The train, of course, is absolutely still,
Or sliding oh so slowly home again,
Bouncing its brittle gunfire up and down.
Kitbags wobble: the dark sacks
Of tonight's resurrection-men,
Stencilled with simple names and broken numbers.
Down the corridors black concertinas
Play the floor sideways, out and back:
Facilis descensus Averno. Under my sandals
The shining parallels race on,
Haul our lost luggage into nowhere
Over cinders, torrents of oily stone;
Our stuffed compartments, weary cubicles,
Whistled dolorously downwind
Under the gantries of the blinded Midlands.

A hanging door sucks at a cold platform;
The steam blows a long kiss to the stars.
Brekekekex, koax, koax, croak the porters,
Peremptory, forlorn. Where are we? There?
But our smoky dream-boat punches up the night,
Which smells of battledress, and I crouch,
Wedged over an empty packet of Wills Woodbines.
A pack of cards flicks and whispers;
The light slithers on a helmet's curve.
As the train dances roughly on its points,
Sleepers keep pace with their reflections,
The dull lanterns of their heavy faces
Nodding, shaking between yes and no.
Calmly, as if incurious quite, they ride
This ebb and flow, the endless of away.

There for the Taking

They are still there—how could they fail to be?
The tree loaded with plums, its sage-green fingers
Knuckled on purples, mottles, bobbles of gum,
The pale snail-shells footing the garden wall.
The brass plate by the door at the end of the drive
Polishes itself away summer by summer;
Its shallow letters tell the most perfect strangers
How fiercely a name burnt for a few short seasons.
Now such things must be left to be themselves,
Things which have lapsed, which have fallen out of love,
Out of the nets we cast so lightly about them,
Greeting their simple natures with eyes and fingers.
Signed like the money-box closed on childish pennies
'For Holidays', 'For Rainy Days, Mine and Others',
They are still there, for others, there for the taking:
The frosted glass at the top of the stairs, the doorknob
Angling again for a child to map its contours,
For the hug and share of the new, tremendous secrets.

The Old Bell-board

The old board's not used much: a panel of glass
Up high on the wall, a square of windows
Where the tremblers shake about in their tiny lairs,
Wagging their tongues, red whirling dervishes.
Someone is holding a bell-push down and down,
Threading the house with a cobwebby scrape of sound.
From a numbered room above or below the stairs—
Bedroom One, or Two, or the Breakfast Room—
A lifeline twitches the coverlets askew,
A muscle ticks away at an eyelid's corner.
Who is to fetch and carry, to come and go,
Swishing the air at the twist of a hand or door?
Grandad trims the gas lamp under his breath,
A terrier, cloudy-eyed under rugs of hair,
Is down at his heels; Granny, bobbed with pearls,
Loops her Waterman over her deckled paper;
Evelyn works the kitchen full steam ahead,
Shaving the runner beans to a quick cascade.
The bells can tell you the special smells of things,
The way time picks at the crumbly bones and mortar.
Is it set fair, or storm? They have no lives,
These citizens of a sleepy house and garden:
This doll's house, weather-house, where grown-ups do
The things that a child imagines they have to do.
Their work is a special kind of play and feasting
As they busy away at their sunshine, letters and dust,
Building and filling the ark that I keep in my head.
It rings like the long note held by a rubbing finger
On the rim of a glass, the rim of a memory.
Those tremblers waver and slow, going nowhere,
Coming so lightly to rest, the current warm in the cell.

At the Stairhead

Your feet could not get colder than this,
Your head hung over the stairwell, the carpet-rods
Pure slips of gold gone into powdery dark.
The shining of things goes down, and always down;
The bed is marooned by acres of crackling board.
Tonight the house is bandaged like a mummy,
But sharp with edges of reflected light.
Waiting, here at the stairhead, crowded pictures
Char on the walls: cold ashes, distant voices.
You grope for them; they are proper to the night,
Voices from catacombs, bone Roman voices,
Voices chinking tipsily in their cups
Or lipping softly down the corridor,
Growling in dullness, rising, slipping, falling:
Kennelled things. Your ears sing with blood,
Alert as if the stars could make frost music
Or bats shake silver chains out of black thickets.
The voices too are making a bad song
Where light might live behind the closing of doors.
There is nothing here for you, just the white cold
Pressing itself on flesh, on the gold slips
Which travel down and down to the powdery dark,
To the nothings of velvet, the stupid voices
In a long house of thin glass and throaty curtains.

Magic Lantern

The magic lantern show is nearly over.
The bilious frog in his tasselled smoking-cap
Croaks and goggles fubsily in his box;
Ageing monkeys, faced in gold and scarlet,
Twirl themselves off with their barrel-organs.
Goodnight to monocles and waxed mustachios,
Priggers and prancers of the Victorian order.
Light now draws perfect family circles,
Grey as smoke-wreaths, dim as muddled homework:
My mother, profound and sleepy as a doll,
Grandad, young and bowlered, slipping himself a grin,
Granny laughing, a dog—wild shreds of hair,
Long nose, black eyes come from the dust
To see nothing that once saw something.
Now for the tailpiece: Grandad's fingers
Twitch past a small black train, mad as a spider,
Tiptoe across a bridge across a ravine
Across green tumbling across to a blue river.
The train pulls gruffly out of Lincoln Central;
My eyes choke on coal dust, search the wind
And clanking signals for some huge delight
Cloudy as last year's pain, bright as a dewdrop.
There is a handle, polished as a nut,
The key to the last slide lifted from the box.
Turn it, the arcs and lozenges dissolve,
Spinning the good dreams and the humpbacked night
To overlapping folds of pink and violet,
The Rose window high in Lincoln's transept.
Granny, the frog, the little tiptoe train
Open and shut, the dog's blind, lustrous eyes
Bright as the eyes of spiderwebs and flowers,
As the next lucky thing hoped or counted for.
The magic lantern is hot silver tin
And if you touch it, it will burn your fingers.

In the Boxroom

The papers are going their special kind of yellow
Up in the boxroom, which is made of dust,
Half-light, and dust, where all the deadlines follow
Their tide-marks back into the cool, undistant past.
The closeness of things is what the boxroom knows,

Keeping its out-of-season Christmas baubles
In cardboard a little too high without a chair,
Locked up with an odd summer or two, and fribbles
Fingers have not yet put to waste or fire:
Hatboxes closed on a twist of empty tissue

Which could be what spiders hang in the high angles
Where the corner line and the ceiling lines converge.
The air dull, but holy. My skin tingles
And the small space is large enough, as large
As its reserve, its kept and unkept secrets

Waiting for nothing, as I stand there watching,
Because they have had their world, as in their time:
The blue yacht with its sails set, the points switching
The ghost of a clockwork train, that thickening ream
Of newsprint faded and dry. I riffle it through

And the *Royal Oak* and the *Graf Spee* go down,
The Old Codgers write dead letters to the dead,
And I pull the headlines back, line upon line,
'The Navy's here!' A hatch battens its lid
Down on the hold of a prison-ship, the *Altmark*.

All the pale mouths, packed under metal rafters,
Will break into soundless cheers, delivered to air,
The boxroom floating loose on the dusky waters,
Glory catching at dust, and the door ajar—
And the house waiting to take me into the garden.

Taking the Paper to Mr Elvis

Toes and fingers touch a perfect circle,
The sun drops gently into outstretched hands,
The blue sky fits the garden like a hat.

God is idle up there, watching the sparrows,
Which never fall; summer a huge clock,
Its heartbeat steadying the unstruck hours.

The gardener's barrow rooted to the earth,
The croquet balls left fruiting on the lawn—
This is the seventh day, on which He rested.

I cross the world, watched by my old familiars,
Clutching yesterday's paper, neatly folded:
News from nowhere, which must be handed on.

The tangled hedge butts sheer against my path;
Quickly, I slip the paper past its teeth.
Someone I do not want to know will come there,

An Otherness, to pry with ancient fingers,
Picking through leaves for my propitiation,
A name as awkward as a foreign country,

At home with different birds, the different green
Of vegetable-things glimpsed between crevices.
I watch my steps back for our preservation,

Counting a held breath, feeling my skin cool,
The sun twisting the light to disbelief,
The garden breaking round me like a glass.

Reading the News

In the long summer light, before bed, when the rooks
Caw, caw, caw over the huge elms,
In a time of kitchen rituals, closing flowers

And the putting away of childish things,
Two old men sit down side by side,
Benched with other, older justicers:

Sibs and gossips who are weatherwise,
Who know the worth and market price of things,
Familiar with their worlds, flesh and devils.

My grandfather adjusts his gold spectacles.
He has made notes on the six o'clock news
Which he reads carefully to the gardener,

Halting his way through acres of grey water,
Deserts and snows, distance upon distance,
Where armies grope and falter, turned away

From old red bricks, green lawns and hedges
Soaked in a generous and golden light.
The sofa-back is cool; I am in church,

Idle at evening service, when large matters
Are turned over in words black as this earth:
Lincolnshire loam. *Caw, caw,* say the rooks.

A Handful of Cards

You, gone under ground, speak kindly
  Out of lost seasons.
They made their round: flower, seed-head, flower;
  Your box hedges

Still fringe their different garden,
  Down sudden alleys
Say what they once meant: hour, day, hour,
  Time for the taking.

In a handful of cards you come to light,
  Your fingers shaping
The simplest words: love, kisses, love,
  'Bother old Hitler!'

Letters to post: your walnut bureau,
  The blotter's mirror,
Closed on the ghosts of our affection.
  Over our shoulders

Time is sacking all red-letter days,
  Blurred postmarks,
His push-bike ticking love, kisses, love.
  The inks are drying

Where the light plays on pale honeysuckle,
  Michaelmas daisies,
A Christmas rose: hour, day, hour,
  The first snowdrops.

'When are you coming?' A handful of cards
  Asks its one question.
Here, bees are homing: flower, seed-head, flower,
  The scent of box leaves.

Blackpool, by Night

'This is a real photograph.' Above,
The lemon moon swings through a gloss of brown
And spangly wings are just this kind of blue
In *Fairy Wedding Tableau No. 2*
When star-tipped wands light up the bridal gown.
'One more for your album. Lots of love.'
This is a real photograph. The sun
Is creasing shadows in my garden hat
And flattening my lifted drawing-book
Which tells your watching Brownie how to look
At Blackpool: bold, and stary-white, and flat.
But loops of wild illumination run
    Out of that glass and paintbox, flowing free
    From pasteboard filmed and cloudy as the sea.

*

Sincere Wishes on Your Birthday

Yes, they were happy as I must have been,
Those model children *Bright as Sunny Hours*
In Fair Isle pullovers and buttoned shoes,
Dandling a bi-plane, reading Mother Goose,
Flashing a wrist-watch—*Gifts in Welcome Showers*—
Each arm as fawn and smooth as Ovaltine.
Such lucky charmers, stuck in lands of then,
Who promised sunbeams, joy and trellised bowers
In which the future might be comforted.
Their noble collie rears a sager head
To swags of buttered stuff; the garden flowers
Prink out their sharp cerise, viridian,
    Re-touch the birthdays: all that brown cortège
    Which travels backwards on the album's page.

Bigness on the Side of Good

I slip down alleys to our other house,
Where Granny Puff jollies us into laughter
And a million snails live in the wet ditches.
I swing on walls there, crouch in a Ford's skeleton
Bunkered by long grass, while all about me
Striped cinnabars are munching up the ragwort.
My Uncle keeps a monstrous pig, an airship,
As pink as Churchill woofling in his den,
All chubby-chopped, rooting for Victory.
My Uncle Tommy is a man you must look up to—
A Captain Flint, the proper size for uncles.
His broad back can take a deck of cousins
Giggling and swaying through a summer day;
Churchill gives his famous vulgar V,
Swelling his voice into a buttery grumble.
I take his name apart: the rich grey church
Set heavily upon its smooth green hill.
My Uncle's pig smells rich and rotten-ripe
As I drape its backside with a Union Jack
While cousins fall about with lunacy,
Shelved on the wall like Beatrix Potter's kittens.
Later, we find the flag gone very missing,
Marked, gorged upon, and inwardly digested;
Our stomachs have to swell with pride or tighten.
The pig died, surfeited by patriotism:
John Bull and the pig, wrapped in the flag or round it.
My Grandad is reciting 'Barbara Frietchie'.
'Shoot, if you must, this old grey head,
But spare your country's flag!' she said.
A pig, of course, would not care to respect it.
My Uncle is the most famous man in Rasen;
He does not know that I have killed his pig.
Everyone greets him as we walk together
To the Observer Post out on the Racecourse
Where German bombers in black hang from the ceiling.
Having fought right through the first World War
He'd know the right way up to run a flag,
How to cure hams, how to bring home the bacon.

Jigsaw

'Off Valparaiso': the full-rigged clipper
Dances out of the dark, her box of sails
Catching the sun upon a difficult sea,
Jimp waves that will not flow, but break stiffly
Against odd patches of this cold sea-bed
Which is the floor of the tray, floor of the table
Where awkward, absent things go back to brown
Out of white wings and glamorous ancientries.
My lead ships tumble from the black playbox;
Greyly, in yellowish light, they gather names:
Ajax, Achilles, Exeter—floating out
Between the headlines and the photographs
Of battle-damage and the cheering dock-side.
Bronzed and fit on my Lusitania medal
A bony Death staffs the ticket-office,
Sending the idle rich to watery graves,
And snug in the glass-fronted bookcase
Boy Cornwell crouches by his dying gun.
Off Valparaiso. I pick up chunks of ocean,
Twisting the sandy plywood to blue water,
Hunting a skyline, picking at horizons,
Popping the bladderwrack between my fingers,
Sniffing green slimes on a sunk breakwater.
The beach is staked and spiked: a scroll of wire
Rubs its sharp back up against the clouds
Sliding its box of sails over the coast,
Off Sutton, Mablethorpe, and Valparaiso.

The Salamander

They were right to wake me, right as Cellini's father
Who saw a salamander dancing in the fire
And boxed young Benvenuto's ears to sharpen memory.
My father can declaim Jean Ingelow
With all the little deaths that live in cadence:
   'Cusha! Cusha! Cusha!' calling,
   Ere the early dews were falling . . .
For when it is High Tide on the Coast of Lincolnshire
The sea goes cusha cusha in my ears,
Pouring slowly through the bungalow curtains.
That night, they woke me. The grey murmur
That sleeps in dry shells on a mantelpiece
Had grown to a mad zoo of tangled furies,
So I was wrapped up, carried to the Prom,
And while the salt air spun my dreams away
Watched slabs of ocean shudder into concrete,
Toss their exploding puffballs at the stars
As all the winds and waters of the world
Sang with the dark, and with the strangeness
Of being the wrong thing at the wrong time and place.
On other nights they could have drawn the curtains,
Opened my bedroom windows to the sky
And shown me live things dancing in the flames,
But, saved from memory by an inch of sleep,
The wafer of a door, a paper blackout,
Those little sands of time that slouch in sandbags,
They let the room throb to its dull migraine,
Things blind and restless roam beyond the pale.
   That flow strewed wrecks about the grass,
   That ebbe swept out the flocks to sea . . .
The poem, too, was something to remember.
I hang upon the last note of the siren,
Feel the last bomber snoring into silence
Till only my blood sings on its one pure note,
Though I hold the room to my ear like a cold shell.

Woodlands, 7 Kilnwell Road, Market Rasen

You carry yourself away: a freight of silence
Closed by the settling dust, the curtains' slow, calm fall
Across your bent glass and the deep lustres,
The phlox, the hot paths and the cherry wall.

Your gas lamps burn on, pressing their yellow blurs,
Their soft snake-hiss against the swarming night,
Your furniture is mirrors, dark smells, lairs;
Lawns level off, your trees twist out of sight—

And all is Lemuria: a family of ghosts.
Manes exite paterni—and thrice again to make up nine.
At any midnight, hands washed, spitting the black beans,
How should I conjure to redeem myself and mine,

Cleanse your barrow of bricks, that scatter of years
Which holds the otherness of once familiar things?
No placation can ease my head, or yours,
Of the dead we knew, and are: the uncrossable rings

Draw tight about us. There, inviolable,
Childhood is sealed off in its rock tomb,
Valley of Kings, rose-red city, long trouble
Of gold masks, love, lost stuff heaped in a locked room.

Blackout

Surely the light from the house must stay at the glass
And pull itself back, into the room again?
For the light is ribbons, faces; it cannot pass
Out into the loose garden, and the rain
Which scrats at the pane with absent, occasional fingers.

And all our drapes, curtains, tackings of cloth
Closed on these glances, glimmers, openings,
Where the cold glass sticks tight to the hung moth
Quivering its white thorax and plumed wings—
How brief they are, how pale those lookers-in—

Are teaching the light it has nowhere else to go
But round the angles of things, filling the square
Where a sofa thickens its back, the bookshelf row
Scatters in flakes of gold, the arm of a chair
Wobbles where shine slinks off, and into the corner.

I slip round the door and the years, into the dark,
My feet risking the gravel, the wind and wet
Slight on my face, but the house is shut: an ark
Where no parting or pinhole eases to let
My eyes build back the room where I am sitting,

Curled on a cushion, dragging the hours to bed,
As I stand outside in the heaviness of the night,
The child in the chair in the room turning his head
To the out that is bonded out from his cage of light.
I am hunting still for a way to return his glances,

For the house is a black cloud stopping the lawn
From sheering off into further wastes of grey;
But the child inside knows well that the blinds are drawn
Not to keep the light in, but the dark away—
And the future, its blind head nuzzled against the window.

GOOD FLYING DAYS

Good Flying Days

Good flying days: the kites kite-shaped
Stirring restlessly in children's fingers.
That urgent, papery flap of trapped insects,
Light laths crossed on whispering cellophane,
And wind bowling over the hard ribs
Where the low-tide forest bares its teeth
Out of the old coal-measures.
Under the gallantries of a flying circus
The usual dog eats up enormous nowheres,
Malcolm Campbell on the Bonneville salt-flats,
And little shrieks bob ankle-deep in foam.
Out of our canvas roll we build a bird,
The crush of silk tightened, polished cane
Slipped into brass ferrules. O monstrous crow,
Hawk of the World, outface, outsoar
The low-sky gaggle, lime and lemon dancers
Tossing their heads, fluttering Chinese pigtails.
As afternoon is polished off to grey
I send my twist of paper up the line
To bring the dark wings down, dismantle summer.
Only air whirring in cloth, sea hissing,
And all the shadows pitching in the grass.
New tides of light bleach out our markers,
Those jaunty signs that we were there, yes, there,
Holding the brief sun on a dipping leash
While water cleaned our footprints from the world.

Going Out: Lancasters, 1944

'They're going out', she said.
　Together we watch them go,
The dark crossed on the dusk,
　The slow slide overhead

And the garden growing cold,
　Flowers bent into grey,
The fields of earth and sky
　Losing their strength to hold

The common lights of day
　Which warm our faces still.
Feet in the rustling grass
　We watch them pass away,

The heavy web of sound
　Catching at her throat.
We stand there hand-in-hand,
　Our steady, shifting ground

Spreading itself to sand,
　The crisp and shining sea.
Wave upon wave they go,
　And we stand hand-in-hand,

The slow slide overhead,
　Stitched on a roll of air,
As if they knew the way,
　As if they were not dead.

Pathfinder

Night-riders gather, and all skies look east.
The stars are steady in their tight formations,
Crossing with light this grind of troubled air.
The house is softening its velvet textures,
Buoyed to its moorings by a bombers' moon.
Its charts record only the stretch of wood
Which gives, misgives, at the sleepwalker's touch,
And stranger-faces turn from watery mirrors.
Outside, the beaten grass is lost and grey
Where levelled meadows launch their thunders out,
Over the shining waters into Europe.
The house is dreaming little flocks of questions:
Fidgets of dark glass tripping in the frame,
Easings of tread and riser at the stair.
The cows at dusk spoke only country matters
As daylight sank behind the window-bars,
And no pathfinder drops his marker-flares
Upon this crouching city of bad dreams.
Over the leaded grate, farmer and dog
Exchange their painted sentiments of loss
As the old home comes underneath the hammer.
The night dies on till all the tides are turned,
And though the ebb was deeper than the full,
From this, our common ground, we cannot single
That diminution in the homing waves
Which speaks of tears, which speaks of tears and flame.

The Loss

Broken out of cold sea, warmish land,
   These tremendous holds of smoky water:
A mackerel sky splaying its bobbles after
   A lather of cumulus; somewhere else a hand
Patching a Dutchman's trousers, just enough blue
   Is showing through

To let me take these hours into a meadow,
   The amber sun floating upon my back,
Prickling small sweat. Turning the other cheek,
   I can hold light enough to cast a shadow
And watch the slight scars travelling overhead:
   The complex dead

In Heaven, where angels high, from rose clouds
   Someone falls the full height of his ambition,
And great fleets, softly plumed, in tight echelon,
   Pass and re-pass: harsh migrant birds,
The grass quivering under their glanced wings.
   Through all their vapourings,

Their strainings at the leash, I pass them on,
   Follow close their dull score of sound.
Manna: the frost of shrapnel on the ground;
   Magnificat in a butterfly-bomb, incendiary fin.
And out of sight, the loss: the children downed
   In the black playground,

The dead reckoning—stars in loose formations
   Taking each bomber up, the iced wings wheeling
Over and out, the slow climb past the ceiling,
   The crew stiff at their seven stations.
Arcturus, Vega: a far field thick with light,
   And the unleavened night.

Jacob's Ladder

When searching cones of light
   Transfix you at your station,
And the radiance foretells
   How night will claw you down
Into a soundless glow,

Who do you see on Jacob's ladder
   When the rungs are blazing,
The children rise in ash,
   Sifting the dark with faces?
Non Angli, sed Angeli,

Pitched from seas of flame
   Into such absences,
Tossed as Dante's lovers
   In cyclones of desire:
Brands plucked from the burning

Quenched in a great cold.
   For flesh packed in cellars,
Bonded, swollen, molten,
   Shrunken out of substance,
Whose hands make cradles?

When the air is full of names
   Blowing in the slipstream,
The city lies in cinders,
   The moon, immaculate,
Crooks her sharpest horn,

Who do you pass on Jacob's ladder
   When the rungs are blazing?
What traffic have the dead,
   Tongues burned, eyes blinded,
As the squadrons turn for home?

Searchlights

God, who struck open skies, unleashed his fingers,
    Making a stab at justice; violet light
Fused the low cloud to his scorched-earth policies.
    I watched him at his random bolts and fireballs,
Then saw, under the skies that had healed over,
    Layering dark upon dark, dark upon dark,
Such towers of white air leaning against the stars,
    That rational light, as pure as gospel-truth,
Seemed to make a clean break for its freedom,
    And whirled over soft ribs of glass
A host of moths, dry silver, Christmas dust,
    Glittered in their endless, quiet fury.

Here, the generator hums the night to sleep
    In a small field, damp with shadows,
But splayed out by a canal in Germany
    Those quiet rods, angling in black water,
Gather to lash a cockpit into radiance
    And wings crumple under their instruction.
'I'm afraid we've had it. I shall have to leave you now.
    Baling out. Good luck, everybody.'
Metals and flesh falling, flakes of ash
    Sifting their way into a later harvest,
His words hang in the increasing silence,
    Uncancelled by those ministers of fire.

A Parachute

Not flying, but floating
As if to land lightly, seed oneself, rise again:
  A sycamore key twisting
  Into its grass nest, life run
    To all that brown

Under all that blue, breeding
Its ash leaves, tufts of grey light, thistledown,
  Each spider laddering
  Its perdurable cord, sun
    Still burning down

These flowers, their heads breaking
Out into blue wings, white wings, pinks flapped
  From a grey bed, silks taking
  Each skirl of wind: dropped,
    The catch slipped.

And a parachute is falling:
A scallop of pale flesh, open-eyed, rose of death—
  Crete, Arnheim. Baling
  Out, out, under the flight-path
    Goes the ghost of a moth,

Ghost-moth, a soul visiting
The mansions of the dead, loss, a dark habitation?
  Only a child's handkerchief, twirling
  String on a dull stone,
    Or lead man

Finding his feet, slipping
Quickly through garden green, against cloud, caught
  On a plum-branch, dropping
  A little further into night,
    The crumpled sheets

Where a child sleeping
Free-falls his limbs, landscapes, his here and there;
   Takes his bearing
   By dream and star,
      And chooses air

   As one unharnessed, leaping,
Span from his plane through night, forest, snowfalls,
   Passed out and through, spidering
   His unsupported miles,
      And woke with men, not angels.

German Bricks

The children's German bricks are boxed up tight
With all their paper citizens. 'Sleep tight.'
A thinnish moon will give the night its head,
Cities and dreams composed under a lid

Of scattered cloud, small hours and the high cold
Whose passengers work hand in glove with cold.
A garden closes on its quiet places;
A boy and girl set out their palaces

Where light chops at water, windows open
On darkish rooms as flowers, faces open,
Fingers lock in fingers over wet sand.
The longest journey starts at the day's end.

And Montaigne reads in his tower; below, the garden.
Night and its cupboard loves confirm the garden:
That dream and harvest eager to placate
The crooked coastline and the bloody mart

Where common things, children, bricks, boxes, tell
The common secrets that they always tell.
A slow cascade of marker-flares goes down;
Clouds of black angels dance upon a pin.

And not a column, brick or spire is lost;
The picture still confirms it: none is lost.
Where Jacob Schutz sheltered, a child cried endlessly
Jesus, my Jesus, Mercy, Mercy.

Service

Hearing the organ stray,
Wambling slow time away
In sit and stand,
A candle-bracket cold to hand,

Watching the shadows pass
Under this burning-glass:
One bright eye
Of lapis lazuli

Over an old saint's head,
A running maze of lead.
Robes of blood,
And little understood

But the high lancet's blue
Which makes the whole tale true
And is definite,
Shifting its weight of light

As afternoon lies dying,
The trebles following,
Stone gone dim,
God down to his last hymn.

AND LITTLE WARS

And Little Wars

We fought the war through
　In small back gardens,
Set on our cloths of gold
　Each makeshift squadron,

Into action-stations
　By the German bricks
Sent our kilted Scots
　With bayonets fixed,

Pressed into service
　Busbied Guardsmen,
Stretcher-case, bearer,
　Machine-gun section,

Dinky half-tracks,
　Churchill, Sherman,
Red-cross nurses,
　Men and half-men,

Searchlight battery,
　Plastic airmen,
Life Guards on horses,
　Crimean cannon.

But too much marching
　With arms shouldered,
Too many sentries
　With arms grounded:

Non-combatants all,
　Swinging their lead
Into our slow-march
　Victory Parade.

War Games

The armies are parading on the carpet.
Spilled from their call-up papers, coloured boxes,
They master dressings, drill-books, about-faces.
Peace was Christmas ribbons, *light wing'd toys*;
War a sad teddy-bear who feeds on dust
And fixes bayonet eyes upon the soldiers.

When I inspect the long files of my soldiers,
As lined and threadbare as this rag of carpet
Whose corners fold themselves upon the dust,
I cannot see into those empty boxes,
These antique fables nor these fairy toys,
But search for something missing in their faces

And search again, for all familiar faces
Are blotted out by rank on rank of soldiers
And icy winds which cry *All is but toys*,
Lifting the corners of the playroom carpet
To show me in the boxroom with the boxes,
Watching the tiny parachutes of dust

Marching a leaden army through that dust
Towards an enemy with equal faces
As dull as mile on mile of cardboard boxes.
Love glitters on the lances of my soldiers,
My carpet knights who dye into the carpet,
Whose brief Commander knows that *dreams are toys*.

Host upon host; *silence you airy toys*,
Wasting through trenches, mountains, into dust.
Somewhere beyond the fringes of the carpet
The tin men put new flesh upon their faces
And swing down bitter streets, playing at soldiers,
Where money rattles in collection boxes.

Nest upon nest of empty Chinese boxes:
Triumphs for nothing and lamenting toys—
Such knowledge fills the nodding skulls of soldiers
On their brief march from ashes into dust.
The past is only what their future faces
Across moth-eaten rolls of shabby carpet.

The coffin-boxes close on precious dust,
And all our toys have names and special faces.
The soldiers are the figure in my carpet.

The Stain

As the stain spreads, working itself coarsely
Into the grain, I watch the colours run,
The fabric weaken, an inky cloudscape loosely
Darken the white cloths and dowse the sun,
Put all the flowers out, fold up the butterflies.

Out in the dead garden I feel it seeping
Through rough grass; the ladybird on her stalk
Knows her house is on fire, neighbours are shaping
Zeros of sound—those bibles of careless talk
Which cost a lifetime to close on their revelations.

It is reaching nearer and further: a salt tongue lapping
The first beach head of sleep; a ghost of grey
Is nibbling at paint, finding our level, keeping
Our small world company, stretching out each day,
Hunting like a bad smell for an unused corner.

The sky absorbs it: the earth, the stripped hedges,
The cold playground, the shiny bedroom floor,
A game of Lexicon, a sampler of brass badges.
In the beginning was the word; the word was war,
And when the word peeled off, skin shrank from air.

A slow recovery, as if a restless sickroom
Left leaves and faces caught in a long half-light.
The time-bomb ticking away under house and home,
I will them back, sealing them under night,
Blacking their eyes and setting their bricks trembling

Under the siren's call, the mumbled anger;
The handfuls of metal rain on a dark street
Whose crackling tarmac fissures into danger,
And all this later sunlight will not make sweet
An ambivalent sky crouched over the new, bright strangers.

Countries

The countries that we made lived in the gardens,
And sat out in the rain, and went to nothing.
The sentries dropped their guard, and watched in silence
As little creatures ate there, and were eaten.
Such countries grew more blurry than the mazes.

Our mazes lived with careful maps of islands,
Their names more fanciful, their space less peopled.
They sprouted palaces on torn-out pages
Which lived and died in schoolrooms: cells of boredom—
As boring, nearly, as a Hampshire barracks.

The barracks, later, where I played at soldiers,
Had nothing much to say about the gardens.
The metal lockers, tall and green, seemed coffins
In which all childhood was dead and buried,
Although its chapter-house comes in the story.

The story I am telling is of landscape:
A landscape words feel moderately at ease in,
As grave as Poussin, or as Auden's limestone,
A pastoral with tombs, and somewhere, soldiers.
The scenery is flat, and grass, and sentries,

Though sentries now are bits of brick and rubble:
The straight and crumbly edges of a jigsaw,
They stand by hangars stamped with blackish numbers.
I drive more slowly when I pass the flatlands,
These uncompleted circles, Nissen shelters.

These shelters hold a scruff of straw and fert-bags,
Puddles of cold and rather oily water.
I try to make out faces in the darkness
And follow gleams of light I think are truthful,
Talking fitfully of wings and runways.

The runways veer across the maps and mazes;
The by-roads break them. By the gates of farmers,
Or where a slab of names, a stuck propeller,
Cries 'Halt!' with the abrasion of a sentry,
I stand and watch the criss and cross of grasses

Which blow, of course, to mirage: green and silver.
Still, if I have a mirage, that is something,
And a big sky holds room enough to write on,
To say that I invent what I remember
Then touch down lightly on this patch of concrete:

Concrete which shivers into cracks and sparkles,
Something left over, out at night, forgotten,
The runways weathered into desert islands
Which suffer the erosion of their coastlines—
Countries over which I stand as sentry.

7 Newton Road, Harston, Cambridge

Your corners jab out at the wind
Punching across dark, levelled land,
A spill of couch-grass, twitch and vetch
Choking the runnels of the ditch
Which pins your garden to a field
Sullen, cloddish, brown and bald.
Your tingling fence-wire clips the sun,
Drying its nets against a pen
Of hens, dull layabouts, who itch
Feathers and grass to a bare patch.
You live with ribbing: knitted clay
And con-trails laid against a sky
Blown out and scuffed until the blue
Rubs off to nothing showing through.
As Doodlebugs come popping on
You work our fingers to your bone,
Your skins grown callous to the hand
In endless making-do and mend:
Eggs drowned in buckets, dusty coal,
A punctured push-bike's ticking wheel.
Perched on the edge of here or there
You stay unhealing: red and raw.
The village runs you out of life
But the harsh air won't scrape you off
Your patch of sand, or make them go,
The sad and mad who neighbour you
And slew you sideways on their pain:
Cold, obdurate, quotidian.
I lie in bed and hear the groan
Of a green pump still working on,
Hauling the overflowing cess
Out of its pit down scrawny grass:
The sour brown gulps, the biting wind,
Dragging you back into the land.

Neighbour

She stands at the fence and calls me till I come,
Mouthing a message for the wind to lip-read;
Only my name stands in the air between us.
A slattern house, reeking of dirt and music,
A compound of long grass, drawn round with wire,
A summer sky, aching with Cambridge blue—
These are the substance of her misery
Which forces through our bonded bricks and mortar.
Her son comes home on leave; the sobbing furies
Thicken at night. Our lives become a part
Of that insistent voice, those wavering grasses.
Her garden burns, the red rim eating out
A heart of black which blows to feathery ash.
Low overhead the puddering doodlebugs
Comb the cold air, each weight of random pain
Clenched tight—these trivial ganglions on the nerves
Of a slow-dying war. My mother's love
Betrays her to the ambulance, and asylum.
All the doors close, but in the silences
Her message stays. There is no way to read it.

Doodlebug (Kirschkern): 1944

Kirschkern, cherry-stone,
  On a blind date,
Counting it out
  To the rim of the plate.

Cherry-ripe, cherry-ripe,
  Ripe I cry,
Who plays at cherry-pit,
  Eats cherry-pie?

Black Heart, Smoky Heart,
  Leaves yellowing,
Sunken cankers, cracked bark,
  A late withering.

One sour Morello
  At the garden wall:
Earth raked over
  Where the silence fell.

Cherry-pip, Kirschkern,
  The plate licked clean.
Who asked for bread?
  Who got a stone?

'Ich hatte einen Kameraden'

'*Ich hatte einen Kameraden* . . .'—the class shuffles;
A tune rises and falls. What did you do?
Went out, looked at the days, clocks handing us on,
And blowing about our heads that grey and blue
Turning to cold, biting through worn gloves.
Kicked at hushes of leaves by a long wall,
Picked up bits of the world and put them down,
Pulled the trees about to make shadows fall,
Drew our bows at a venture from Drift to Lane,
From Quarry to Field, from there to back again.

And then? Begged at the door, lay out on the lawn
Feeling the eager differences of green,
Watched stick-insects doing nothing, slowly,
Picked the circling bones of the seasons clean.
Went in, cutting wet rind from our shoes,
Fingers fondling Trudi the dachshund's head;
Lit by diamond panes and painted china,
Salted the butter softly on to the bread.
Set out the glories, the servitudes of war
In close formations on a cockled floor.

And what did you say? Nothing, nothing at all.
The wind and the rain have scrambled our words deep
Into a huge silence. Our mouths opened;
Secrets were exchanged. They are kept, and keep
For ever, ever. Do you want a name?
It is the name of your own once-special friend.
'*Einen besser'n find'st du nicht*' the class wails on,
Locked in iron-bound desks till the song must end,
Till the scrape and blur, the chattering out of sight—
Our faces garbled, strange, swung out of the light.

The Old Gang

The Reichsmarschal is pouring through the window,
Puff-pastry, lardy-cake: his tunic creases
The crazy-paving on the shroud of milk
Hung out to dry upon my bedtime mug.
His breeches swell with meat, his polished jackboots
Suck their black liquorice up against his thighs.
He puckers his face into a little smile,
Tricking love-crinkles from his currant eyes
And littering the night with iron crosses.
Oh wicked uncle in your fancy-dress,
Tinkling a cataract of cold regalia,
I know you have the shadows for your meiny.
Their liquefaction stiffens to salute;
You break your jewelled baton into music,
The taratantara of Roman trumpets
Banging my head against a moving staircase
Of light and thunder. Now, in his night-school,
Old Gob comes at me with his wooden face,
His long cane swishing at my flying heels,
His long gown scrumpled over secret fire.
The Honour boards glister in heavy gold:
Famous forgotten names floating on oil.
The fat man and the thin man meld their hungers
In one enticing and protracted smile.
The Reichsmarschal smacks his buttered lips;
Old Gob mounts the dais, holds his rally down.
The dream roars, and all the hands fly up:
Hurrah, hurrah! *Sieg heil, Sieg heil!*
The bedroom window has three iron bars
On which I know my knuckles clench and whiten,
Suffering again this tongueless lashing,
Vaulting the wooden nightmares in their stables.

The Wooden-heads

I am learning nothing at incredible speed,
Weltering in ink: a blotted copy-book,
My nibs and fingers crossed unhopefully.
Crumbled from the soft pages of a library book,
The wooden-heads are punishing my daydreams.
Zombie-skittles, they run rings round you;
The circle closed, the victim vanishes.
Having cleared London, they will start on Cambridge.
Ropes and wall-bars offer no escape-routes;
Matched at bantam-weight, I bob and weave,
My gloved hands pummelling the air.
They have caught Hodges; by the cycle racks
I saw him, red-faced, simple, yellow-haired:
Now empty, strangled in his father's toolshed.
The foul-mouthed kapos of the middle school
Are doing a roaring trade in small extinctions,
Trawling a dragnet for us small-fry.
The game is gas chambers. Into the labs with you,
Heads pressed hard against the bunsen burners
Whispering you to sleep. We lurk in crevices,
Closed by dark wood, secrets, grey flannel,
A master with one eye like a scorched Cyclops,
And feel our little strategies go down,
Slipping through all the circles of the underworld.
The gowns and blazers form their broken ring;
There is a gap still I can wriggle through
If I can only purge away my scraps
Of dog-Latin, wrong notes, false equations.
Best to go to ground in the reeking bogs:
Unbolted stalls, chads, pig-troughs of urine.
The wooden-heads are out looking for minds to bend,
My satchel of trash blocked up in their beast-world.
I see Hodges swinging behind a door,
His childhood flushed away, his bright blue eyes
Still laughing at me over his knotted throat.

His Face

His face is everywhere: a slab of paste,
Dabbed forelock, hot pale eyes. His lair
Is a wild nest of burnt metal, shadows,
But I know a good-luck works between my fingers
In Granny's brooch, the soft red-leather Kipling
Where his black-magic cross stands gold and upright
Or swings its level arms over its shoulders.
Its firework cartwheels on his wooden sleeve
Into the pages of a million schoolbooks
As he strikes air rigid with his quivering fingers,
Brooding impossible rages with a tongue
That skids unbraking over the crackled ether.
His flat hand launches out his throbbing arrows,
Flung through classroom dust and the long nights
When scuttle helmets bob in rows before him,
The banners dip, the oily torches flare,
The high goose-steppers kick the air aside.
Tiring, he twists his fingers in his mouth,
Blows out his brains, dowses himself in petrol,
Blazes himself away, runs wild with Eva, Blondi,
All the cold wolves who stuck to him like glue
And slavered to pull down the sun and moon.
Now, he shrinks to a well-practised doodle,
A dead thing which thrills my poking fingers.
When I peer at the cracks between the clouds
I know his fleets will rise to his command,
Their crushed bones hauled and straightened from the tomb,
The cowlicked skulls perched in their greenhouse cabins
Blinding their way across the sulky ocean.
In the School Dining Hall I take my Spam
And cut it to a neat, pink swastika.
I eat it slowly; it is his communion.

The Finger

It is the accusing finger, always saying:
'I denounce, and you are guilty, guilty
Of having eyes, a nose, of being you.'
The grey-haired Hausfrau clutches at her chair,
Her elbows bent; Herr Nobody is crouched,
His face a blank space where the room writes fear.
Between the two of them, a wireless-set
Crunches the moment up between its teeth,
Babbling its careless talk, costing their lives,
And at the door, their child, a bantam rooster,
Perched, like me, on the edge of double-figures,
Stands blazing: damned and righteous
In little shorts, an armband, open shirt.
Behind his finger the Gestapo tower,
Putting on their high hats, drawing guns.
                                                I know him:
This little swimmer into cleanness leaping,
Hiker, boxer, dipped by his brown heels
Into that lake whose baptism will wash
Blonder than blond, whose duty is betrayal.
I look long at his eyes, his accusation
Where all dark sulks, feet kicking at stones,
Pounding fists, tears in cold bedrooms
Flower into self-possession. I too accuse
The cat, the flowers, and the foolish clock
Ticking itself away in dwindling circles.
Blood-brother, wait for me, and I will bring
My cub-cap, toggle, green garter-flashes,
Walk with you, horrid friend, make a stick-camp,
Swap stories and insignia: gleefully
Explore the boundaries of our pride and shame.

The Shiny Pictures

The shiny pictures go straight through my eyes,
But nothing in them adds up to a view.
I feel that I must push those leaves away,
That if I stare enough I'll come to know
The secrets of the road, the blurry trees

And that small figure staring straight ahead,
Dressed in the black and white of photograph,
Walking quite briskly; nothing in his face
To tell me what on earth he's thinking of.
Tin-helmeted and holstered at his side

Two soldiers elbow him like special friends,
The kind who chew gum, warn you out of harm.
I watch them lead him to the proper place
Then level off their rifles, take their aim
And squeeze their gentle triggers with big hands.

His body loose, cords crossed upon his chest,
Earth gazing greyly at his nodded head,
His blindfold white, his jacket creased on blood—
This is how people go to being dead:
Their glasses taken off, backed-up to posts.

Quisling, Fifth-Columnist . . . the sense of it
Is hidden in what papers always say;
But that's not much about the bits of cloud,
The nothingness, cold happenings, more sky.
I turn back to the start, and let the flat

And boring ground, the figures, trees and road
Go sliding by to their appointed ends,
Which are no ends, but dark re-surfacings
Of eyes, and guns, and gentle-seeming hands
From thick old pages, heaviness like lead.

The Word: 1945

The word is a word stuck behind closed lips.
I take these pictures carefully to bed,

Finger my skull, shape the familiar room
Into some kind of space: a yard, huts.

I can make no sense of these deep pits,
Branches of flesh and bone, once green and quick,

This guesswork of smudged breasts and genitalia
Speaking greyly of guilt in classroom whispers.

I lie there with the word the house won't say;
The pictures move about inside my head.

I cannot fit the word into the war.
I look down quietly at my striped pyjamas.

Souvenirs

Something enormous, with a molten core
Of huge delight: here is a bit of shell,
A whorl of light I picked out of the gutter
Where last night's gunfire did its giant leap,
Stumbled and fell in thousands on the city.
Brass cartridge-cases wild as buttercups,
A bomb snug in long grass under blossom,
Strips of tinfoil stuck on eastern hedgerows—
Autolycus could stuff his ditty-bag
With all these little deaths of shot and shell.
Granny tacks up my shoulder-flashes, chevrons:
A reliquary cased in brass and velvet
Where regiments glow fiercely in my bedroom—
Spiced, musty fragments dedicate to Mars,
Sparkling with war, war, war—a word
As dark and rich as last year's postage stamps
Heavy in orange, red and cobalt-blue.
Old nights grow thick as blackness cramped in boxes,
Kept for something, hoarded; briefly opened
Into cloud split and sharpened by pale sunlight,
Days bright and stiff with names and battle-honours.
The souvenirs lie truthfully in their box,
Cold-hearted now, and most inscrutable,
Dead as John Dee's obsidian scrying-stone.
I know unguessable things are stirring there,
Troubling with their fumes and cloudiness
A past as deep and shrouded as the future,
Where planes fly backward onto broken runways
And stamps released from dried and creaky hinges
Fly round the jarring world with urgent letters
Addressed to strangers, crooked heaps of rubble.
I watch my ghosts stand to their glowing anvils,
Beating their ploughshares back to swords again.

The Field Transmitter

The Field Transmitter, heavy in its box,
Uncurls its oily braids and hanks of wire,
Stuck by a green corrosion to brass terminals.
The knurled knob taps and stutters dit and dah,
Its V for Victory: 'For you, the war is over.
Come in, my children, from the echoing green,
The city street as yet unlicked by paint;
Climb from the bunkers in your sad back-gardens,
Yesterday's foxholes: iron, sacking, iron.
Hand in your outsize helmets, bits of perspex,
Your bomb-fins and that treasured German arm-band;
Dismiss those leaden armies to the dust
Which settles into what you will call memory.
Crouched for the last time on the garage floor,
Let my headset's hard constriction tighten
Till all your war becomes a new, strange tinnitus,
The bombers climbing through your cloudy brain-cells,
Gaining their altitude and levelling off
In as much sky as spreads from ear to ear.
This band of gunfire bouncing off your skull
Makes the thick sound of other children dying,
Out of your reach, beyond your messages,
Who played their war-games, heard the sirens glow
Hot silver filaments in miles of night,
Till gathering babel took them to its arms
And held them still, and held them very still.'

V. E. Day

Carpamus dulcia: nostrum est
Quod vivis: cinis, et manes, et fabula fies.
PERSIUS: Sat. V

Noticing oddly how flags had been rubbed thin,
Bleaching in shut drawers, now unrolled
In blues, reds, their creases of old skin
Tacked on brown lances, headed with soft gold.
    Clotheslines of bunting,

And light fresh at the front door, May
Switching the sky with stray bits of green,
The road levelling off; the day much like a day
Others could be, and others might have been.
    A woman laughing,

Sewing threadbare cotton to windy air,
The house open: hands, curtains leaning out
To the same gravel, the same anywhere, everywhere.
Birds remain birds, cats cats, messing about
    In the back garden.

And a table-land of toys to be put away,
To wither and shrivel back to Homeric names.
Scraps gathering myth and rust, the special day
Moving to its special close: columnar flames
    Down to a village bonfire

In which things seasoned and unseasoned burn
Through their black storeys, and the mild night
Fuels the same fires with the same unconcern:
Dresden, Ilium, London: the witch-light
    Bright on a ring of children.

Night, and the huge bombers lying cold to touch,
The bomb-bays empty under the perspex skull.
The pyres chill, that ate so fiercely, and so much,
The flags out heavily: the stripes charcoal, dull.
    Ashes, ghosts, fables.

The Other Side of the Hill

Who is it mooning about in his half-lit room,
Sharing my name, the set of my bones: a boy
Hung in my chains of words, puzzling it out
Under suspended judgements, gagged and blind
While Time poisons him cloudily under the door.
What are these thoughts I badger him into thinking,
These possible routes I map for him, joining the dots,
Holding his secret writing up to my fire?
Oh, he is poor Jim Jay, stuck fast in Yesterday,
With his party tricks, the clumsy deck of snapshots
He shuffles, cuts and deals, those blind fingers
For what I have dropped and lost on his bedroom floor.
I would like him to put our heads together, confer
About old hedges, pavements, the dazzling film
Which is racing out of sync with its dusty soundtrack.
Perhaps, though, all he can say is 'I was there',
As if 'there' were a special place, or a right of way,
Not a sentence shared with the other prisoners:
The sad Italians behind the bulldog gate
In the flat, unhurried fields, who turn and wave,
Ridiculous in their Commedia diamonds, moons,
Our next-door neighbour, jabbering lost Polish,
The bitter stench of her house dark in our nostrils.
She paces for ever behind her shining fence-wires
Which sing as I pluck them: the Chekhovian sound
Of a nerve thrilling under a huge dull wound.
Absorbed, at home in this peculiar country,
He leans from my Tower of Babel, anxious to help me,
Then settles back again into his thousands,
Swinging his satchel, counting his souvenirs,
Reading the signs on the other side of the hill.
For old time's sake I offer him this future:
Another chance to make things up. After all,
I am the victim in his puzzle-picture;
He is the child that I go looking for.

Under the Barrage

Schlafe, mein Kind
In your mother's bed
Under the barrage.
The soon to be dead

Will pass you over;
It's not your turn.
The sheets are warm
But they will not burn.

In your half-dream
Sirens will sing
Lullay, lullay,
Lullay my liking.

A saucer of water
For candlestick:
The night-light steadies
A crumpled wick

In a house of cards
In a ring of flame;
The wind is addressed
With a different name.

Schlafe, mein Kind
In your mother's bed.
Under the barrage
The soon to be dead

Lie as you lie,
And will lie on
When the dream, the flame
And the night are gone.

OXFORD POETS